Microbial Pathogen Survival and Extraction From Point-of-Use Filters

Awwa Research Foundation

About the Awwa Research Foundation

The Awwa Research Foundation (AwwaRF) is a member-supported, international, nonprofit organization that sponsors research to enable water utilities, public health agencies, and other professionals to provide safe and affordable drinking water to consumers.

The Foundation's mission is to advance the science of water to improve the quality of life. To achieve this mission, the Foundation sponsors studies on all aspects of drinking water, including supply and resources, treatment, monitoring and analysis, distribution, management, and health effects. Funding for research is provided primarily by subscription payments from approximately 1,000 utilities, consulting firms, and manufacturers in North America and abroad. Additional funding comes from collaborative partnerships with other national and international organizations, allowing for resources to be leveraged, expertise to be shared, and broad-based knowledge to be developed and disseminated. Government funding serves as a third source of research dollars.

From its headquarters in Denver, Colorado, the Foundation's staff directs and supports the efforts of more than 800 volunteers who serve on the board of trustees and various committees. These volunteers represent many facets of the water industry, and contribute their expertise to select and monitor research studies that benefit the entire drinking water community.

The results of research are disseminated through a number of channels, including reports, the Web site, conferences, and periodicals.

For subscribers, the Foundation serves as a cooperative program in which water suppliers unite to pool their resources. By applying Foundation research findings, these water suppliers can save substantial costs and stay on the leading edge of drinking water science and technology. Since its inception, AwwaRF has supplied the water community with more than $300 million in applied research.

More information about the Foundation and how to become a subscriber is available on the Web at **www.awwarf.org.**

Microbial Pathogen Survival and Extraction From Point-of-Use Filters

Prepared by:
Geoffrey B. Smith and **Kevin H. Oshima***
Biology Department
New Mexico State University, Las Cruces, NM 88003

*Current Address: U.S. EPA, Cincinnati, Ohio

Jointly sponsored by:
Awwa Research Foundation
6666 West Quincy Avenue, Denver, CO 80235-3098
and
U.S. Environmental Protection Agency
Washington D.C.

Published by:

Awwa Research Foundation

American Water Works Association

IWA Publishing

DISCLAIMER

This study was jointly funded by the Awwa Research Foundation (AwwaRF) and the U.S. Environmental Protection Agency (USEPA) under Cooperative Agreement No. X6-830948-01. AwwaRF and USEPA assume no responsibility for the content of the research in this publication of the opinions or statements of fact expressed in the report. The mention of trade names for commercial products does not represent or imply the approval or endorsement of AwwaRF or USEPA. This report is presented solely for informational purposes.

CONTENTS

TABLES

FIGURES

FOREWORD

The Awwa Research Foundation is a nonprofit corporation that is dedicated to the implementation of a research effort to help utilities respond to regulatory requirements and traditional high-priority concerns of the industry. The research agenda is developed through a process of consultation with subscribers and drinking water professionals. Under the umbrella of a Strategic Research Plan, the Research Advisory Council prioritizes the suggested projects based upon current and future needs, applicability, and past work; the recommendations are forwarded to the Board of Trustees for final selection. The foundation also sponsors research projects through the unsolicited proposal process; the Collaborative Research, Research Applications, and Tailored Collaboration programs; and various joint research efforts with organizations such as the U.S. Environmental Protection Agency, the U.S. Bureau of Reclamation, and the Association of California Water Agencies.

This publication is a result of one of these sponsored studies, and it is hoped that its findings will be applied in communities throughout the world. The following report serves not only as a means of communicating the results of the water industry's centralized research program but also as a tool to enlist the further support of the nonmember utilities and individuals.

Projects are managed closely form their inception to the final report by the foundation's staff and large cadre of volunteers who willingly contribute their time and expertise. The foundation serves a planning and management function and awards contracts to other institutions such as water utilities, universities, and engineering firms. The funding for this research effort comes primarily from the Subscription Program, through which water utilities subscribe to the research program and make an annual payment proportionate to the volume of water they deliver and consultants and manufacturers subscribe based on their annual billings. The program offers a cost-effective and fair method for funding research in the public interest.

A broad spectrum of water supply issues is addressed by the foundation's research agenda: resources, treatment and operations, distribution and storage, water quality and analysis, toxicology, economics, and management. The ultimate purpose of the coordinated effort is to assist water suppliers to provide the highest possible quality of water economically and reliably. The true benefits are realized when the results are implemented at the utility level. The foundation's trustees are pleased to offer this publication as a contribution toward that end.

Walter J. Bishop
Chair, Board of Trustees
Awwa Research Foundation

Robert Renner, P.E.
Executive Director
Awwa Research Foundation

ACKNOWLEDGMENTS

We would like to acknowledge the professional and efficient oversight of Dr. Misha Hasan of AwwaRF. She has been very helpful at various important junctures of the study. The advise and editorial critiques of the Project Advisory Committee [Drs. Anne Camper (Montana State University), Greg Harrington (University of Wisconsin), Dennis Juranek (Centers for Disease Control), and Tim Straub (Battelle Pacific Northwest Labs)] has been helpful throughout and is much appreciated. We are indebted to Ms. Marget Foutre of the Water Quality Association (WQA) and Ms. Anita Kantak of Applied Research-West, Inc. for allowing us to participate in the development of the National Consumer Water Quality Survey questionnaire. We would also like to acknowledge the helpful advice from Mr. Joe Harrison of the WQA.

Mr. Joe Leslie of Barneby and Sutcliffe, Inc. was helpful in identifying the source of activated carbon used in this study, and Mr. Guy Giordanengo and Mr. Given Clatterbaugh of the American Tank Company were very helpful in providing market-level information regarding the particular model of the PUR™ carbon block filter to be used in this project.

Lastly, we would like to acknowledge the excellent New Mexico State University students who collected much of the data for this project: Ms. Jennifer Walsh, M.S., Mr. Carlos Melendez, M.S., Ms. Rachael Hamel, B.S., Ms. Katie Souza, B.S., Mr. Brian Oase, Ms. Ayshea Quintana, Ms. Varissa Benally, and Ms. Carmela Smith.

EXECUTIVE SUMMARY

The foundational idea for this project is that household faucet-mounted water filters may be used as bioforensic sampling devices to detect the extent of a bioagent release. The aims of this project were threefold: 1. Determine the most popular point-of-use (POU) filter used in the U.S. for use in the current study. 2. Develop an efficient method to extract potential bacterial and viral pathogens from these filters. 3. Document the extent of pathogen retention that occurs in POU devices, and the effect that POU filter age has on microbial recovery efficiency. In collaboration with the Water Quality Association, a questionnaire was modified that was used in a national survey (WQA, 2004), and from the reported percentages of U.S. households which use water filters, it is estimated that 32 million households utilize faucet-mounted water filters. The survey identified PUR™ faucet-mounted water filters as the most commonly used POU filters used in the U.S. According to representatives of American Tank Company, the only U.S. distributor of the entire line of PUR products, the PUR Ultimate™ carbon block filter is their top-selling PUR filter, and so this was the filter chosen for the majority of the work reported here. Brita On-tap™ filters were also tested on a limited basis since Brita filters had close to the same market share as PUR filters. Since more than 30 million U.S. homes use faucet-mount water filters, the idea of utilizing these filters to sample for a potential bioagent release is feasible in regard to the number of these types of filters in U.S. homes.

A systematic study was conducted to determine the best method for desorption of microorganisms from powdered activated carbon (PAC) alone, and these results were used to determine if carbon block filters can be used in bioforensic applications. The following test organisms were used in different applications of the study: *Escherichia coli* K12, *Pseudomonas aeruginosa*, *Bacillus subtilis* spores, *Salmonella typhimurium*, and the enteric bacteriophages PP7 and MS2.. The *Bacillus* spores. *E. coli* and PP7 virus were exposed to powdered activated carbon sources and subsequently were eluted using both chemical (surfactants, proteinaceous compounds, charged compounds, and varying pH) and physical (blending speed and extractant pressure) variables. Microbes were quantified by culture-based techniques. Microbes adhered strongly to PAC, with an average of 79% of inoculant bacteria attaching, and 94% of inoculant virus attaching to PAC. Proteinacous, surfactant and ionic-based extractants were tested as to their ability to desorb microbes from PAC, and it was found that proteinaceous compounds and surfactants extracted the most microbes when used together. Typically, bacterial extraction efficiency from PAC was significantly higher (percent of *E. coli* extracted ranged from 3.9% to 37.8% of those attached) than viral extraction efficiency (less than 1% was desorbed from PAC by extractants).

Carbon block filters (either 1-stage or 3-stage PUR Ultimate™ filters) retained greater that 99% of spore and vegetative forms of bacteria, whereas virus were retained to a lesser degree by the filters (for PP7 virus, 76 % $\pm$ 8% were retained on 1-stage filters and 92% $\pm$ 5% were retained on 3-stage PUR filters (+/- S.D.). However, viral extraction from carbon block PUR filters was much higher than from PAC, averaging typically 10% of the virus attached, and bacterial extraction ranged from 10 – 20% extraction efficiency. The most efficient extractant was identified as a combination of 1% peptone and 1% Tween-80, and filters were back-flushed at 0.5L/min using this extractant. Microbial extraction of bacteria, spores and virus from another top-selling water filter (the Brita On-tap™ filter) was very similar (14% extraction efficiency) to the PUR filters (15% extraction efficiency), indicating that the proposed extraction procedure is functional and comparable with the two top-selling faucet-mounted filters. When

PUR filters were aged to varying degrees (new, 20%, 50% and 100% of the recommended lifetime), extraction efficiency of virus, vegetative and sporulated bacteria (ca. 15%) was largely unaffected.

In mock release studies in which filters were challenged with test bioagents and then exposed to tap water flows and frequency-of-use typical of U.S. households, *B. subtilis* spores could be consistently extracted from filters, and were detected four hours, 2 days and 5 days after bioagent release. When tap water was spiked with 10^2 spores/mL, greater than 99% of the spores were retained by the filters, and the percent of attached spores that were recovered ranged from 10.4% at day 0, to 4.3% five days after the release (averaged from five separate experiments). In contrast, *E. coli, Salmonella typhimurium* and a representative enteric virus (PP7), were rapidly inactivated in the chlorinated tap water, indicating their improbable function in bioagent releases in chlorinated water supplies. Of course in water supplies that are not chlorinated or have attenuated chlorine levels, these agents may survive distribution. In two mock-release trials, it was demonstrated that filters exposed to bioagents could be refrigerated for 3 days and bioagent spores were still detectable, thus making the approach more field- practical. The extraction method developed here was capable of extracting bacterial and viral agents from two of the top-selling faucet-mount filters, albeit with less than 50% extraction efficiency. It is concluded that household water filters can be used as microbial sampling devices for bioforensic applications in the event of a bioagent release into the U.S. drinking water supply.

Future work is needed to increase the sensitivity of the detection assays, to develop a molecular method that functions in the presence of activated carbon residue, and to test more types of filters in water with, and particularly without, chlorine.

CHAPTER 1: INTRODUCTION

NEED FOR THE RESEARCH; STUDY OBJECTIVES

The U.S. drinking water supply remains a vulnerable target for potential bioagent release, and, among numerous other preventative biodefense issues, there is a need to identify the origin and extent of a bioagent release should it occur. In the aftermath of a bioagent release, it will prove extremely difficult to retro-actively sample enough of our water supply to gauge the origin and extent of the release. In the American Water Works Association Research Foundation (AwwaRF) request for proposals for this grant, the very interesting idea of using household drinking water filters as devices to potentially monitor bioagent release event was proposed. Two of the most commonly used POU filters are manufactured by PUR and Brita (see results of national consumer survey in Chapter 3, section I, below), and both the faucet-mounted POU devices utilize carbon block filters with a one um pore-size rating, such that greater than 99% of protozoa are trapped in these filters (Watertanks, 2006). With the pore-size of the carbon block, it can be expected that a high percentage of bacteria will also be retained in these types of filters. In a recent U.S. national survey (Water Quality Association, 2004, discussed below), 28% of households have some type of water filtration product installed in the home, and the highest percentage of these (39%) are faucet-mounted filters. Using Census Bureau estimates of the U.S population in 2004 (293.6 million, U.S. Census Bureau, 2006) this represents 32 million U.S. households which have faucet-mounted filters. Therefore, the use of faucet-mounted water filters as in-situ microbial sampling devices could span, in time and location, potential domestic water supply bioagent releases, and microbial sampling from these filters could provide a meaningful "record" of any such release.

Therefore, the current study was designed to examine some of the practical aspects of this idea of utilizing household water filters as bioforensic devices. Representative bacterial, viral and protozoan agents were used in this study to represent the three most important classes of microbes that potentially may be used as bioterrorist agents. The study had the following objectives.

Objective 1. Analysis of national surveys to identify the most common POU devices currently used in U.S. households.

Objective 2. Identify the most efficient, highest recovery method to extract and enumerate representative bacterial and viral agents, then test this method's recovery efficiency with representative bacterial and viral agents.

Objective 3. Document the extent of pathogen retention that occurs in POU devices, and the effect that POU device age has on retention and recovery of pathogens

POU FILTRATION EFFECTS ON BACTERIAL HETEROTROPHS AND PATHOGENS

Point-of-Use (POU) filters for tap water are frequently used in household applications to remove odors and organic chemicals. Most POU devices utilize activated carbon as the main adsorbent material because of its very high specific surface area and the presence of many different types of surface functional groups that become sites for adsorption (Han et al. 2000). Studies have shown that POU filters can trap microorganisms, but may also increase the number of heterotrophic bacteria in effluent water compared to systems without POU systems. Wallis et al. (1974) documented the occurrence of higher levels of bacteria in POU treated water compared to untreated water, particularly after periods of inactivity. Heterotrophs have been documented to similarly increase in the effluent of POU filter devices after periods of inactivity (Bell et al. 1984). Studies have also shown that pathogen retention and survival in POU systems are greatly decreased when the filters are already colonized by heterotrophs (LeChevalier and McFeters. 1985; Camper et al. 1985b; Reasoner et al. 1987). Recently, Chaidez and Gerba (2004) have reported that water, after POU filtration, had elevated levels of opportunistic pathogens such as *Pseudomonas aeruginosa*, and *Aeromonas hydrophila*, compared to untreated tap water.

METHODS TO EXTRACT BACTERIA FROM ACTIVATED CARBON

Activated carbon is a common treatment component of POU systems and potentially may be an effective concentrator of pathogens. Three protocols are summarized that were designed to extract bacteria from granular activated carbon (GAC) material (Camper et al. 1985b, Stewart et al. 1990 and Snyder et al. 1995). Camper et al. published a paper evaluating a number of procedures to extract bacteria from granular activated carbon (1985a), and then used the most optimal technique to document the survival of pathogens in GAC filters (1985b). Though the Camper et al. (1985a) work was a significant advancement in bacterial extraction techniques, the interpretation of the results is limited due to the highly variable GAC material that was used. Similar variability issues were encountered in the current project. One significant finding reported by Camper et al. (1985a) was that coliform bacteria were killed during sonication even at the minimum settings while the sample was kept on ice.

Stewart et al. (1990) modified the procedure slightly by using 10-fold less peptone in the extraction buffer. Snyder et al. (1995) modified the Camper (1985b) procedure by sampling commercial POU devices using a sterilized no. 6 tree borer, and instead of sampling the entire length of the filter, they removed two samples, from the top and bottom 5 cm of the filter.

One must be careful in using detergents to desorb microbial cells from solid surfaces because detergents can lyse cells, particularly charged detergents (Prescott et al. 2002). In the GAC extraction protocol used by Camper et al. (1985b) and followed by Stewart (1990) and Snyder et al. (1995), the zwitterionic detergent, Zwittergent 3-12 was used. It is important to note that this is a quaternary ammonium detergent because quaternary detergents are most commonly used in disinfectants due to the fact that they disrupt microbial cells and denature proteins (Prescott et al. 2002). Though Camper et al. (1985b) used only 1 uM of this detergent, quaternary ammonium compounds are extremely potent disinfectants: it took 280 times less quaternary ammonium compound to kill Gram-positive and Gram-negative bacteria than phenol (Prescott et al. 2002). Nevertheless, high recoveries were reported by Camper et al., indicating little or no toxicity due to their extractant. To summarize, the biggest challenge in extracting cells off solid surfaces is to develop a protocol aggressive enough for effective desorption, yet mild enough to avoid cell lysis.

In a study by Garduno et al. (2002), it was found that cells of the pathogen *Legionella pneumophila* were susceptible to lysis by the anionic detergent SDS, but were at least 200-fold more resistant to the non-ionic detergent, Triton X-100. However, Gauthier et al. (2000) showed that even concentrations as low as 0.1% Triton X-100 caused the release of both membrane and cytoplasmic proteins from pathogenic strains of *E. coli*. One of the efforts in this proposal was to screen non-ionic detergents for the characteristic of effective desorption without cell lysis.

CHAPTER 2: MATERIAL AND METHODS

MODIFICATION OF WQA NATIONAL SURVEY TO INCLUDE IDENTIFICATION OF TOP-SELLING BRAND OF POU FILTER

In collaboration with Ms. Marget Foutre of the Water Quality Association (WQA) and Ms. Anita Kantak of Applied Research-West, Inc. (and with help from Mr. Joe Harrison of WQA), two questions were added to the national water quality survey that WQA was sponsoring. (Water Quality Association, 2004) The additional questions were the following: 1. What type of (water filtration) product did you purchase? This was asked of people who indicated they had a water filtration product. The second question that was added was: 2. "What brand of water filter was purchased?" The effort that went into getting these two questions added to the survey was considerable!

The report presents the findings of a national telephone survey conducted by Applied Research-West, Inc. between February 15-February 28, 2004. A total of 2,004 adults over the age of 18 and living in private households were interviewed. Applied Research-West used a random sampling procedure and the survey results are accurate within +/-2 percent. The results are representative of all U.S. adults over 18.

MICROBIAL CULTURES USED

Microbial species used in this study: *Escherichia coli* K12 (NMSU culture collection # 25), *Pseudomonas aeruginosa* (ATCC 15692), *Salmonella typhimurium* (National Animal Disease Center, Ames Iowa, NADC # 798), *Bacillus subtilis* var niger (EPA strain received from G. Rice), bacteriophage PP7 (ATCC 15692-B2) and MS2 (ATCC 15597-B1), and the *Cryptosporidium parvum* (P102, Ames isolate) was purchased from Waterborne, Inc, New Orleans, LA as a 5% formalin-inactivated stock. *Pseudomonas aeruginosa*, *E. coli* K12 and *S. typhimurium* were grown in 3 mL of nutrient broth for 17 hours on a shaker (37° C, 150 rpm) for each experiment. The inocula concentrations used in the experiments varied up to 20% from the intended concentration, but the actual concentrations were quantified by plate counts in each experiment. To purify spores, *B. subtilis* was grown for 6 days (150 rpm at 37° C) in nutrient broth supplemented with $MnCl_2$ (40 mg/L) and $CaCl_2$ (100 mg/L), after which the spores were pasteurized (20 minutes, 70° C) and repeatedly rinsed with sterile cold distilled water according to a protocol provided by V. Hill of the CDC (see SOP Spore1.1 in Appendix A). Spores were stored in 40% ethanol at 4° C. Initially, there was a problem of spore clumping (as identified by microscopy) using this spore preparation method, but we have been able to disaggregate spores by diluting the spores in 0.85% NaCl / 0.1% Tween-80, and vortexing the sample for 2 minutes before use. Bacteriophages PP7 and MS2 were purified according to EPA Method 1602 and references therein. The phage were stored in nutrient broth at 4° C for the working stock (up to 1 month) and at -80° C for prolonged storage.

When target microbes were quantified in the presence of non-sterile conditions (ie., microbes native to tap water), the following methods were employed. *E. coli* and S. *typhimurium* were quantified on the selective, differential media mEndo LES, and Hektoen Enteric agar (HE agar), respectively. For quantifying spores, the samples were pasteurized for 20 minutes at 70°C and then plated out on nutrient agar. In previous work, this pasteurization treatment eliminated 100% of *B. subtilis* vegetative cells (J. Rao, unpublished data). The double agar technique was used as described in EPA method 1602 to quantify PPY and MS2 virus using *P. aeruginosa* and *E. coli* (F-Amp strain) as respective hosts.

MICROBIAL ADSORPTION / DESORPTION FROM PAC

The source of the PAC used in this study was that identified by Joe Lesley (Barneby and Sutcliffe, Inc.) that is used as the carbon block matrix in Procter and Gamble water filters (the manufacturer of the PUR filters used in this study). The PAC (50-200 mesh) is a coconut shell-based material manufactured by Barneby and Sutton that is available from Fisher Scientific (cat. # 05-690B). For comparative purposes, the wood-based PAC from the same company that Camper et al. (1985 a,b) utilized, was also used (Nuchar cat. #SA-20 188-R-04). Adsorption to PAC consisted of inoculating 10^6 / mL of *E. coli* K12, *P. aeruginosa*, or bacteriophage PP7 into a homogenizer tube (Fisher, cat. # 05-562-10K) containing 10 mL of 10 mM MOPS pH 7, and 1 g of PAC was added to each tube and the tubes were placed on a shaker (50 rpm) at room temperature for 30 minutes. All tubes were then centrifuged at a slow speed (1075 x g for 4 minutes) to pellet the PAC. The supernatant was sampled for unadsorbed organisms and then decanted. The carbon pellet containing adsorbed organisms was then used in microbial extraction experiments.

Ten mL of extractant solution in 10 mM MOPS pH 7 (unless otherwise stated) was added to the carbon pellet (from adsorption experiment) in duplicate or triplicate samples. The samples were vortexed and homogenized with an Omni Mixer ES Homogenizer (Omni International, ES115) at a speed of 8000 rpm for 1 minute. An ice bath was used to keep samples at 4° C $\pm$ 2° throughout homogenization. Samples were immediately centrifuged at a slow speed (1075 x g) for 4 minutes and the supernatant was sampled to determine number of organisms desorbed. See Standard Operating Procedure Carbon1.4 in Appendix A for more details.

MICROBIAL ADSORPTION / DESORPTION FROM C BLOCK FILTERS

Before adsorption was performed on new carbon block filters, the filters were wetted with 1 L of sterile distilled water. Lower, more realistic concentrations were used in the filter experiments compared to the PAC experiments. Organisms were inoculated into 3 L of tap water at concentrations of 10^4 virions/mL for PP7 and 10^3 cells/mL for *E. coli* K12, and *B. subtilis* spores (higher concentrations of virus were used to ensure detectability). The inoculum was mixed and an initial sample was taken. The inoculum was then divided into six, 500 mL aliquots and 500 mL was pumped through 6 individual filters at a rate of 500 mL/ min, at a pump flow rate of 500 mL/min. Two filters were challenged at a time. Effluent samples were taken to determine the number of unadsorbed organisms. After the 500 mL inocula was pumped through the filter, air was pumped through to ensure that all 500 mL had been pumped through the filter, and to ensure that there was no measurable volume of inocula left in the pump tubing.

Organisms were desorbed from carbon block filters by back-flushing with 500 mL of extractant at a rate of 500 mL/min. All extractants were buffered with 100 mM MOPS pH 7 unless otherwise stated. Effluent was collected and sampled for number of organisms extracted from the filter. Refer to Figure 2.1 showing the pumping system to adsorb and desorb microbes using the PUR Ultimate™ filters. See also Standard Operating Procedure Filter.1.2 in Appendix A. The volume retained by PUR 3-stage filters after air was pumped through was not considered in the extraction efficiency calculations, nor was possible microbial adherence to surfaces such as tubing. Thus, it is a procedural recovery reported rather than a formal mass balance. Based on the fact that both sorption and extraction steps utilized the same materials and procedures, we expect that the number of sorbed microbes to be directly comparable to the number extracted.

Typical volumes and masses applicable to the system used to sorb and extract microbes form PUR Ultimate™ filters are the following:

Volumes:
1 m of influent tubing = 45 mL
PUR Ultimate™ filter pore volume = 140 mL
1 m of effluent tubing = 45 mL.
Masses (obtained after cutting a PUR Ultimate filter open*
1'st stage fiberglass wrap-around filter = ca. 1 g (estimated)
2'nd stage carbon block filter: 89.3 g
3'rd stage mineral layer = 9.9 g

*Note: manually releasing the carbon block inner filters from either the PUR or the Brita faucet-mounted POU devices takes about 30 minutes per filter and a fair amount of effort, using a hack-saw, screw-driver and hammer. This was another reason why back-flow extraction was chosen as part of the proposed procedure.

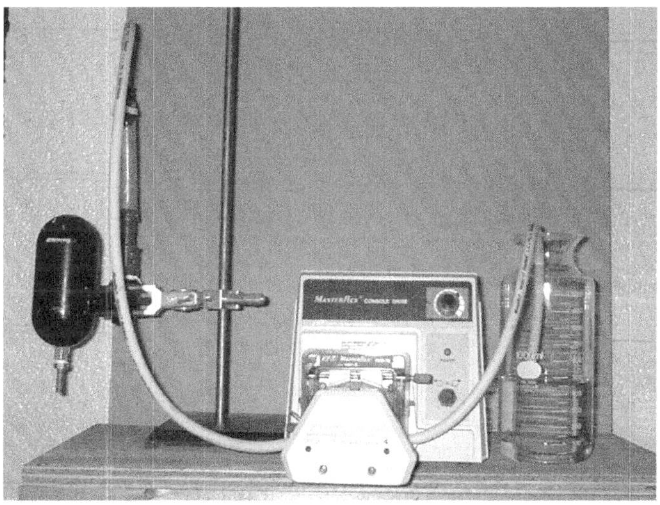

Figure 2.1. Pumping system for sorption/desorption of microbes from PUR Ultimate™ carbon block filters. For desorption, the supply line was cleared and then attached to the bottom of the filter, and the filters were back-flushed with extractant.

USE OF FILTER MANIFOLD TO "AGE" CARBON BLOCK FILTERS AND ITS USE IN BIOAGENT MOCK RELEASE EXPERIMENTS

Filters were aged using a tap water manifold on which up to 12 filters can be installed. The flow frequency was controlled by a timer which was set to run tap water through filters 10 times daily for 5 minute increments on the hour (ie., tap water was pumped through the filters once an hour for 10 hours, and was not used for 14 hours each night. See Figure 2.2 for a photo of the manifold. Individual flow valves were set such that each filter was exposed to tap water at a rate of 0.8 L min^{-1} Both the frequency and the flow rate were set based on typical U.S. household water use (Mayer et al. 1999). A filter was considered 100% aged when 100 gallons passed through the filter, as per the manufacturer's recommendations.

Test organisms were used in a mock microbial release experiment on the filter manifold. Fifteen liters of tap water was inoculated with 10^3 virions/mL of PP7 coliphage and 10^2 cells/mL of *E. coli* K12, *B. subtilis* spores, *S. typhimurium*, or *C. parvum* oocysts. All 15 liters were passed through the 6 filters in a 5 minute time period on day 0. Tap water was run through the filters in the same manner as the filter aging process for 5 days. Two filters were harvested on day 0, 4 hours after inoculation, on day 2, and on day 5. Microbes were extracted from filters using a solution of 1% peptone and 1% Tween-80. Samples of the inoculated, unadsorbed, and desorbed organisms were taken. Samples of the manifold effluent (containing unadsorbed microbes) were taken after the 1st minute, 2nd minute, and 4th minute of filter inoculation as well as the 1st, 2nd, 3rd, and 4th hour after inoculation. Effluent samples (0.5 mL and 0.1 mL) were plated directly onto media, and so detection limits were 2 CFU/mL). In some experiments, two days after the challenge, duplicate filters were extracted immediately, and an additional two filters were stored for three days in a refrigerator at 4°C before extraction.

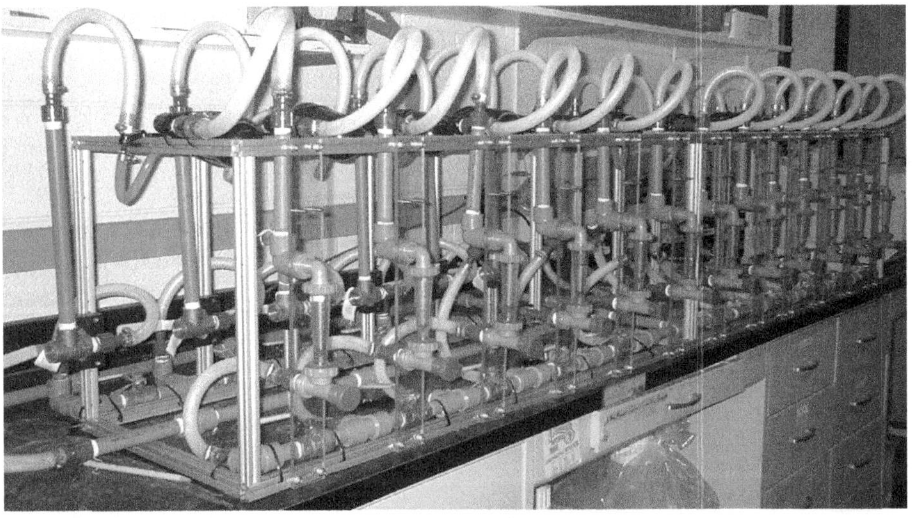

Figure 2.2. Photograph of manifold with 12 PUR Ultimate™ filters mounted. Filters are exposed to tap water at a frequency and flow rate to mimic typical U.S. household use in the kitchen (adapted from report by Mayer et al. 1999).

DOMESTIC WATER SUPPLY CHARACTERISTICS

New Mexico State University tap water was used in this study and shown below are some of its properties (Table 2.1). It is drawn directly from groundwater, is chlorinated and then supplied to the university community. The chlorine levels ranged from 0.03 to 1.6 ppm throughout the study period.

Table 2.1
Representative chemical parameters of the tap water used in this study

Analyte	Results
Total Coliforms	Absent
Turbidity	0 NTU
pH	7.85
Sulfate	250 ppm
Hardness	210 ppm
Alkalinity	160 ppm
Fluoride	0.49 ppm
Total trihalomethanes	0.0258 ppm
Copper	0.015 ppm
Lead	0.0076 ppm

Data obtained from a July 1, 2005 Consumer Confidence Report (NMSU, 2005)

CHAPTER 3: RESULTS AND DISCUSSION

NATIONAL DRINKING WATER SURVEY RESULTS

Most pertinent to the goals of the current project, is the information gathered as to the most common type and brand of POU device identified in the national survey (Water Quality Association, 2004). Shown in Figure 3.1 is the survey question regarding what type of purified water is used in the home, and the resultant responses are graphed. Twenty-eight percent of the respondents indicated that they use some form of water filtration as their method to treat water (Figure 3.1). Figure 3.2 shows the question asked regarding what type of water filtration system is used in the home, and 39% of those having water filters used faucet-mounted filters. Lastly, question 53 of the survey inquired as to what brand of water filter was purchased. By far, the two most popular brands of water filters were PUR (29%) and Brita (27%) (Figure 3.3).

Q. 30. Do you currently do any of the following? **(READ LIST. CHECK ALL THAT APPLY.)**

	Yes	No
Use bottled water	1	2
Boil tap water	1	2
Have any type of water filtration product installed in the home	1	2
Have a water softener installed in the home	1	2
Have a refrigerator with water filter	1	2
Other (Specify):______________________________	1	2

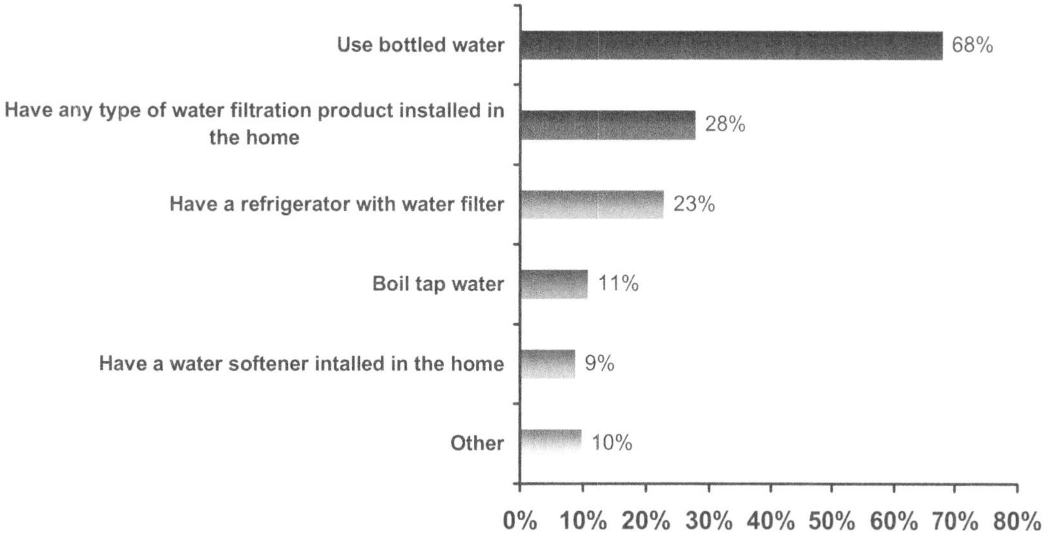

Figure 3.1. WQA national survey question 30 on what type of water treatment is used in the home, and the responses obtained (Source: WQA 2004, National Consumer Water Quality-Research Summary. Applied Research West, Inc. With permission from the Water Quality Association)

Q. 55. [Among those having a water filtration product]: What type of water filtration product do you have?

One that is installed under the sink (i.e., UTS)	1
One that is installed on top of the counter top (i.e., CT)	2
One that is faucet mounted (i.e., FM)	3
One that came with refrigerator	4
Free standing pitcher	5

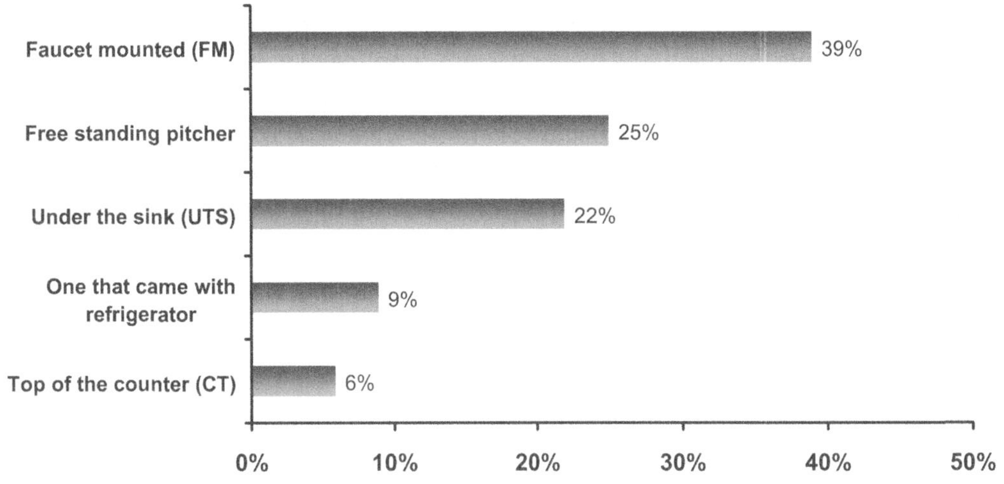

Figure 3.2. WQA survey response to question 55 as to what type of water filtration system is currently used in the home (Source: WQA 2004, National Consumer Water Quality-Research Summary. Applied Research West, Inc. With permission from the Water Quality Association)

53. [Among those having a water filtration product]: What brand is that?

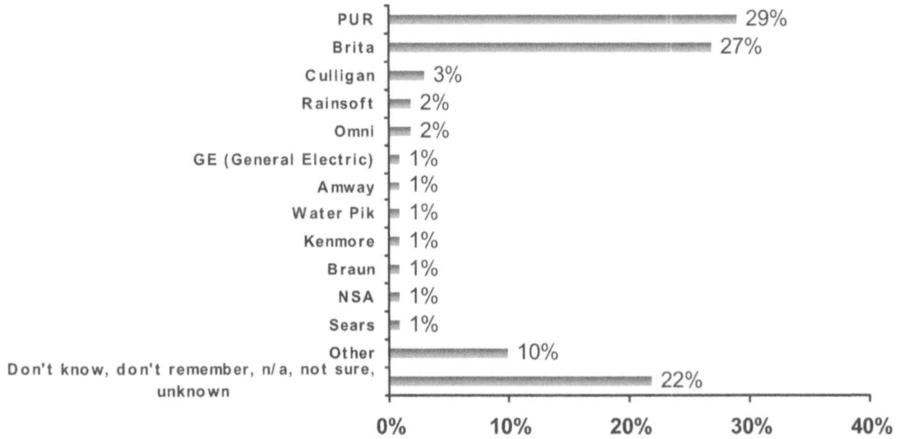

Figure 3.3. Responses to question 53 on what brand of water filter is used in the household (Source: WQA 2004, National Consumer Water Quality-Research Summary. Applied Research West, Inc. With permission from the Water Quality Association)

In summary, using census bureau estimates of the U.S. population in 2004 (293.65 million, U.S. Census Bureau, 2004), the WQA survey indicates that 28 % of the U.S. population (82 million) use some type of home water filtration, and 39% (32 million) of these households utilize faucet-mount filters. Therefore, the idea of using faucet-mount filters to survey for a potential water-borne release of bioagents has merit in terms of the number of U.S. households that utilize these types of POU filters.

With a survey margin of error of 2%, there was no difference in Brita's market share (27%) and PUR's market share (29%) of water filtration products. However, considering that a substantial portion of Brita's market share represents their free-standing pitcher filters, it was decided to use PUR faucet-mount filters for the bulk of the work reported here. Nevertheless, in a few trials, side-by-side experiments have been performed extracting target microbes from both PUR filters and Brita On-Tap™ filters. In order to identify the most commonly used of PUR faucet-mount filters, representatives were contacted of the American Tank Company, Windsor, CA., the only company which distributes all of PUR's products. According to Mr. Guy Giordanengo and Mr. Given Clatterbaugh of American Tank Company, their top-selling PUR faucet-mount filter is the PUR Ultimate™ filter. This, then, was the particular filter used in most of the work presented here (and both 1-stage and 3-stage forms of the PUR filters were compared).

MICROBIAL SORPTION AND EXTRACTION FROM ACTIVATED CARBON

Three types of microbial extraction experiments have been carried out in this study: Extractions from 1. activated carbon, 2. individual carbon block filters and 3. carbon block filters on a tap water-supplied faucet manifold. The percent of microbes recovered from these matrices (or the Percent Extraction Efficiency) is calculated as follows:

[1] No. of attached microbes = No. before filtration – No. unattached after filtration

[2] % Extraction Efficiency = No. of extracted microbes / No. attached microbes x 100

Also note that due to difficulties in extracting and purifying nucleic acids from activated carbon (see details in section below), all the sorption and extraction data described in this report are quantified from culture-based techniques.

Physical and Chemical Parameters Tested Separately.

Both physical and chemical means of extraction were analyzed to identify the optimal method of extraction for bacteria and virus. Unless otherwise specified, the experiments utilized the coconut-based powdered activated carbon (PAC, 50-200 mesh, Fisher cat. # 05-690B), had three replicates per treatment, and each experiment was repeated at least once.

After a 30 minute exposure to PAC, microorganisms adsorbed to PAC at high levels: over 16 separate experiments, 79.4% (Standard Deviation, SD, = 7.5) of approximately 10^7 *E. coli* cells adsorbed to 1 g of PAC, whereas *Pseudomonas. aeruginosa* adsorbed with a mean of 78.5% (14.8) over 8 experiments, and the PP7 virus adsorbed with a mean of 94.1% (5.4) over 7 experiments. Figure 3.4 shows epifluorescent photomicrographs of *E. coli* K-12 adhered to PAC (A, B) and GAC (C).

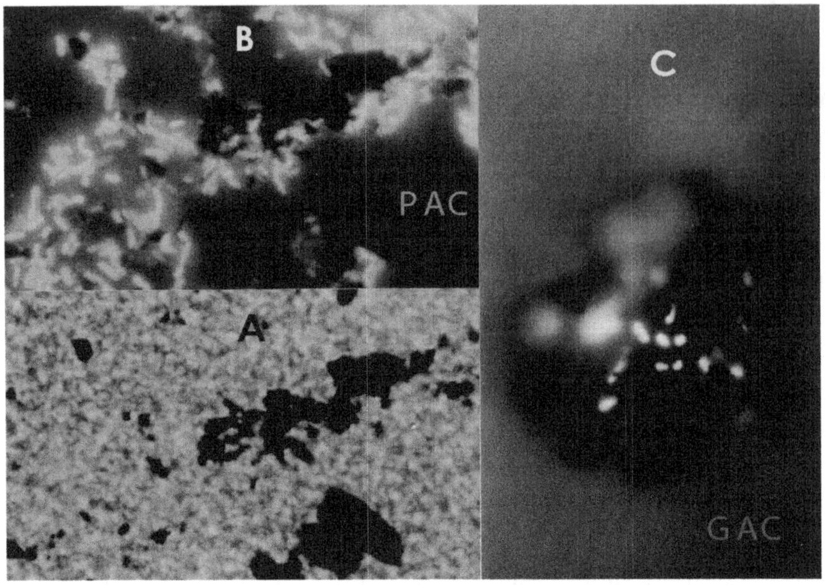

Figure 3.4. ***E. coli*** **strain K-12 adhered to PAC (A and B) and GAC (C) after 10^6/mL bacteria was exposed to 1g of activated carbon for 30 minutes. Photo A shows the PAC (black amorphous material) under light microscopy, and photo B is the same field illuminated by ultraviolet light in which** ***E. coli*** **is visible due to staining by acridine orange. Photo C was illuminated by UV, with acridine orange-stained bacteria adhered to the large granular carbon.**

Numerous chemical extractants were tested (Table 3.1) to desorb bacteria and virus directly from the activated carbon without any physical means (homogenization) of extraction (though samples were shaken at 50 rpm). When tested individually, extraction efficiency (% of those attached to the activated carbon that were extracted) was low, ranging for *E. coli* from 0.01% to 0.15% of the attached bacteria, and for *B. subtilis* spores from 0.03% to 5.4% extracted.

Table 3.1
List of compounds used for desorption of microbes from activated carbon

Proteinaceous compounds	Surfactants	Other
Peptone	Tween 20	MOPS (buffer)
Tryptone	Tween 80	Sodium polyphosphate
Bovine serum	Triton X-100	Sodium pyrophosphate
Bovine serum albumin	n-octyl-beta-D-glucopyranoside	EGTA
Glycine	Glucopyranoside	

Homogenization using an Omni Mixer was the method used to physically extract microbes from activated carbon, and Table 3.2A demonstrates the effect of homogenization speed and duration on extracting *E. coli* from powered activated carbon. It can be seen that extraction efficiency is higher with homogenization, and that 180 seconds at 16,000 rpm gave the highest extraction of *E. coli*. However, when testing the effect of homogenization on *B. subtilis* spore extraction, it was surprising to find that this speed and duration resulted in a lower extraction efficiency than lower speeds and duration (Table 3.2B). In other experiments (data not shown), *E. coli* extraction efficiency at 8000 rpm for 60 seconds was no different than higher speeds and durations, and so the experiments described in the next section were done at this setting to optimize spore extraction.

Table 3.2
Effect of homogenization speed and duration on bacterial extraction efficiency from PAC. Homogenization experiments were performed in the presence of 0.1% peptone (pH 7) and n indicates the number of separate experiments. (Standard deviations are in parentheses)

A. Percent of attached *E. coli* that were extracted from PAC (ie., % Extraction Efficiency).

	90 s	120 s	150 s	180 s
8000 rpm				4.4% (3.6) n=4
12000 rpm				2.8% (1.0) n=3
16000 rpm	7.6% (5.7) n=5	3.9% (5.2) n=2	6.7% n=1	10.5% (6.1) n=6
18000 rpm	5.2% (3.5) n=2			
20000 rpm	4.7% (0.3) n=2			

B. Percent of attached *B. subtilis* spores that were extracted from PAC

	90 s	120 s	150 s	180 s
8000 rpm	2.4% (0.9) n=2	1.8% (1.2) n=2	2.5% (0) n=1	0.6% (0.5) n=3
12000 rpm				0.3% (0.4) n=2
16000 rpm	1.3% (0.6) n=2	0.2% (0) n=2		0.5% (0.7) n=4

Chemical Extractants Tested Simultaneously with Homogenization.

When different chemical extractants were used in combination with physical means of extraction (homogenization at 8000 rpm for 60 sec), recoveries of microbes increased. For example, various proteinaceous compounds were tested (Table 3.3), and bacterial recoveries ranged from 3.9% to 39%. Since recoveries appeared highest in the presence of Peptone, the Tween 20 and 80 surfactants were added to peptone to examine for additive extraction effects (Table 3.4). The addition of the surfactants increased *E. coli* extraction efficiency from 16% to over 30%, and in the case of *P. aeruginosa*, from 19% to over 30%. However, in the presence of both proteinaceous compounds and surfactants, viral recoveries were still less than 1% of those attached to the PAC.

Table 3.3
Percent recovery of adsorbed microorganisms from PAC using proteinaceous compounds

Proteinaceous compounds	PP7	*E. coli* K12	*P. aeruginosa*
5% Peptone	0.24% (0.20)	30.8% (22.7)	19.6% (9.5)
5% Tryptone	0.11% (0.02)	19.0% (8.8)	-
5% BSA	0.08% (0.05)	3.9% (2.1)	-
0.1% Peptone	0.69% (0.50)	29.4% (7.7)	10.4% (1.7)
5% BSA + 0.05 M glycine	0.90% (0.38)	37.8% (7.6)	-

Note. Percentages are means of 2 or 3 separate experiments. A border indicates separate experiments. Standard deviation (in parentheses) between experiments is noted. Dashes indicate that no results exist for that compound.

Table 3.4
Effect of surfactants on microbial desorption from PAC

Surfactants	PP7[a]	*E. coli* K12	*P. aeruginosa*
1% Tween 80	0.10% (0.12)	30.8% (15.6)	29.4% (2.0)
1% Tween 20	0.02% (0.01)	35.7% (9.3)	47.0% (11.6)
Control	0.02% (0.01)	16.4% (3.6)	19.6% (9.5)

Note. All surfactants were in the presence of 5% peptone or 0.1% peptone ([a]) at pH 7. Control was either 5% peptone or 0.1% peptone ([a]). Percentages indicate the mean of 1 experiment for PP7, 2 experiments for *P. aeruginosa* or 3 experiments for *E. coli* K12. Standard deviation is in parentheses.

In an effort to increase viral recovery, the pH of the extractant was varied from pH 4 to pH 9 in the presence of two different concentrations of peptone (Table 3.5). In spite of the5 order of magnitude difference in hydrogen ion concentration, viral recoveries from pAC consistently remained below 1% of those attached. A third experiment in which pH was varied between pH 4 and pH 9 was repeated in the presence of 5% BSA and 0.5 M glycine and, again, very little effect was observed due to pH and recoveries remained low (data not shown).

Table 3.5
Viral (PP7) recovery efficiencies (# PFU Desorbed / # Attached x 100) from PAC as affected by pH in the presence of two concentrations of peptone (0.1% Peptone data, represent the means of 3 experiments, and 5%, the means of 2 experiments)

pH	0.1% Peptone	5% Peptone
	---------- % Virus Extracted	-----------
4	0.04 (0.03)	0.05 (0.05)
7	0.27 (0.14)	0.33 (0.10)
9	0.20 (0.22)	0.28 (0.26)

In similar experiments, Camper et al. (1985a, 1985b) were able to recover more than 83% of *E. coli* attached to PAC using the same or similar homogenization settings and chemical extractants. In an effort to replicate Camper's results, the wood-based PAC (Nuchar catalog # SA-20 188-R-04) was obtained from the same company they used (Nuchar). Four experiments comparing recoveries from the coconut-based PAC to the wood-based PAC were carried out. In three of the four experiments, more *E. coli* was recovered from the coconut-based PAC than the wood-based PAC that Camper used. In the one experiment where we observed better recoveries from the wood-based PAC than the coconut-based PAC, the differences were relatively small (ie.,5.3% recovered vs. 3.0% recovered, respectively.) The mean recoveries of *E. coli* among the four experiments from Camper's wood-based vs. the coconut-based PAC used in this study were 6.7% (8.9) and 21.6 (17.6). Though the experimental variation was high in these experiments, it can be concluded that reason for our lower recoveries compared to Camper's were not due to using a different source of PAC. However, because Camper ground granular activated carbon into an unknown mesh size, we were not able to replicate the exact size of PAC. Additionally, Camper et al. used a river isolate of *E. coli* which we did not have access to. It should be noted that in this PAC-source comparison, 0.1% peptone was used in this study, which gave the highest recoveries reported by Camper (1985a), ie., 83-94% recovery. Other than the difference in PAC size and strain of *E. coli*, we are unable to identify the recoveries reported here from PAC are lower than those reported by Camper et al.

MICROBIAL SORPTION AND EXTRACTION FROM CARBON BLOCK FILTERS

Based on the Water Quality Association national drinking water survey (WQA, 2004), PUR faucet-mounted carbon block filters were identified as the most popular POU devices in the U.S. (see discussion above). According to representatives of American Tank Company, the only distributor of all PUR water filtration products, the PUR Ultimate™ one-stage filter is the best-selling PUR filter. So, 50 one-stage filters were purchased for use in the experiments described below. However, in March when another set of 50 filters were reordered, we received PUR Ultimate™ 3-stage filters with the explanation that PUR had discontinued the one-stage filters. A local survey of commercial distributors verified that the one-stage filters were no longer available.

Microbial Sorption/Desorption from New PUR Ultimate™ and Brita On-Tap™ Filters

Multiple experiments were initially done using new filters to characterize sorption/desorption in the absence of chlorine and tap water microflora. Bacterial retention in both 1-stage and 3-stage carbon block filters was high. When 10^3 CFU/mL was pumped through filters at 0.5L /min, *E. coli* K12 and *B. subtilis* spore retention was $\geq$ 99% of the initial population. In most experiments, no bacteria were found in the effluent sample (i.e. no unadsorbed bacteria were detected). However, PP7 retention was generally less than bacterial retention, and was more variable, as 76.1% $\pm$ 7.9 were retained on 1-stage filters and 92.1% $\pm$ 5.3 on 3-stage filters.

Very interestingly, while usually less than 1% of virus were extracted from powdered activated carbon, viral desorption from either 1- or 3-stage carbon block filters was one to two orders of magnitude higher than from PAC (Table 3.6 and the following experiments described below). *E. coli* K12 desorption from PAC and carbon block filters generally followed the similar extraction efficiency patterns, in multiple experiments, ranging from 3.9% to 37.8% and 7.6% to 32.7%, respectively. For both PP7 and *E. coli*, proteinaceous extractants increased recovery from the carbon block filters compared to using water only as extractant (Table 3.6 and 3.7, and though compounds like peptone were necessary, higher concentrations did not consistently increase recoveries (Table 3.7, and other experiments not shown).

The higher viral recovery from carbon block filters compared to PAC is in large part due to the carbon block design, because the PAC used in the above experiments was the same as that used in PUR filters (Joe Leslie, Barneby and Sutcliffe, personal communication). (Note that Calgon Carbon Corporation has since incorporated Barneby and Sutcliffe, and Proctor and Gamble incorporated the PUR company.) Once the PAC is glued into carbon block, the surface area must be affected relative to viral adsorption sites, since ca. 25% of the virus exposed to the single phase carbon block filters flowed through without being retained. This is in contrast to viral exposure to PAC in which greater than 90% of virus adsorbed almost irreversibly.

In making these comparisons between adsorption to PAC and to C-block filters, the differences in experimental variables should be noted. In the PAC experiments, 10 mL of 10^6 virions or bacteria/mL were exposed to 1 g of PAC, and the 1 g was extracted in 10 mL of extractant buffered with 10 mM MOPS (pH 7). For carbon block filters, 500 mL of 10^4 virions/mL (or 10^3 bacteria/mL) were exposed to the filters, and the ca. 100g of carbon block was desorbed with 500 mL of extractant buffered with 100 mM MOPS (pH 7). Thus, for virus comparisons, 10^7 virions were loaded onto 1g PAC, and 5 x 10^6 virions/100g = 5 x 10^4 virions were loaded per g of carbon block.

Thus, the reason for lower viral adsorption and higher desorption from single phase carbon block filters compared to PAC is not due to over-saturating the carbon block because of a higher # virion / g carbon ratio. There was a 10-fold difference in concentration of MOPS used to buffer the extractant between PAC and carbon block, but later experiments showed little difference in extraction efficiency from carbon block filters due to MOPS concentration (data not shown).

Table 3.6
Extraction of PP7 virus and *E. coli* bacteria from 1-stage (A) or 3-stage (B) carbon block filters (PUR Ultimate™ faucet-mounted filters) as affected by proteinaceous extractants

A. 1-Stage Filters*

Peptone concentrations	PP7	*E. coli* K12
Sterile water	4.3% (0.08)	0.15% (0.04)
3% Peptone	37.3% (22.9)	9.4% (1.6)
0.1% Peptone	6.4% (1.3)	5.5%
1% Peptone	7.2% (2.0)	9.2% (0.73)
5% Peptone	11.6%	10.4% (2.2)

B. 3-stage filters**

Proteinaceous compounds	PP7	*E. coli* K12
1% Nutrient broth	22.4% (27.5)	6.9% (3.1)
1% Peptone	23.6% (17.8)	7.6% (5.2)
5% Bovine serum	2.4% (0.16)	7.7% (4.5)

* For the 1-stage experiments, percentages represent recoveries from a single experiment. Standard deviations (S.D.) are in parentheses, and where not shown, results are from a single filter experiment. Borders in the tables indicate separate experiments.
** For the 3-stage experiments, percentages are the means of 3 experiments (S.D.).

Similar to results from PAC, proteinaceous compounds improved microbial extraction efficiency of both bacteria and virus, and so other compounds such as surfactants were tested in the presence of 1% peptone (Tables 3.7 and 3.8). As was observed with 1-stage filters, , adding peptone and surfactants increased extraction efficiency of bacteria and virus compared to water extraction (Table 3.7). Though the between-experiment variability was fairly high, the addition of surfactant to the peptone extractant increased recoveries of PP7, *E. coli* and *Bacillus* spores. All three surfactants (Tween-80, Triton X-100 and the glycolipid, octyl glucopyranoside) similarly increased extraction efficiency (Tables 3.7 and 3.8). Though octyl glucopyranoside apparently was optimal for PP7 extraction, repeat experiments did not yield near as high of virus extraction (data not shown).

Table 3.7
A single experiment comparing 3-stage filter extraction as affected by peptone and surfactants

Eluent	PP7	*E. coli* K12	*B. subtilis* spores
RO water	3.0% (1.7)	6.3% (3.5)	2.4% (0.6)
1% Peptone + 0.1% Triton X-100	15.4% (5.5)	27.5% (5.3)	6.3% (0.9)
1 % Peptone + 0.1% Tween 80	13.9% (2.0)	23.3% (0.9)	3.5% (3.1)

Table 3.8
Recoveries of microorganisms from 3-stage filters in the presence of 1% peptone with and without surfactants*

Extractant	PP7	*E. coli* K12	*B. subtilis* spores
1% Peptone only	7.8% (1.7)	14.0% (5.3)	3.4% (0.6)
0.1% Tween 80	14.0% (9.5)	18.0% (12.3)	9.5% (4.2)
1% Tween 80	19.3% (21.5)	24.6% (14.5)	20.4% (3.8)
0.1% Triton X-100[a]	15.4% (5.5)	27.5% (5.3)	6.3% (0.9)
5 mM n-octyl-beta-D-glucopyranoside[a]	54.8% (1.4)	8.7% (6.3)	6.7% (7.1)

* Percentages are means of 3 experiments and standard deviations are in parentheses. Borders indicate separate experiments.
[a] Only 1 experiment was conducted. Standard deviations are between 2 replicate filters.

To test whether electrostatic interactions were important in the microbe/filter adherence, charged compounds like polyphosphate were added to the peptone/Tween-80 extractant, or, alternatively, the pH of the extractant was varied (Table 3.9). In three replicated experiments, the only apparent effect of polyphosphate was a decrease in PP7 viral recovery (Table 3.9). In a separate experiment, bacterial extraction was unaffected by pH, though a neutral to basic pH increased PP7 desorption over the use of an acidic pH.

Table 3.9
The effect of charge on microbial desorption from carbon block filters. Three experiments with varying concentrations of sodium polyphosphate were conducted in the presence of 1% peptone and 0.1% Tween 80

Extractant Variable	PP7	*E. coli* K12	*B. subtilis* spores
1% Sodium polyphosphate	6.4% (1.3)	32.7% (17.1)	14.8% (4.2)
0.1% Sodium polyphosphate	9.0% (3.1)	27.7% (12.7)	14.3% (4.7)
0% Sodium polyphosphate	20.1% (2.5)	29.6% (2.6)	12.6% (5.1)
pH 4	5.1% (1.0)	15.7% (0.0)	10.6% (0.29)
pH 7	8.2% (0.2)	15.7% (2.5)	7.1% (6.5)

Note. The percentages in Table 3.10 reflect the mean of 3 experiments and standard deviations are in parentheses. One experiment varying pH in the presence of 1% peptone and 0.1% Tween 80 was performed.

As discussed above, overall sales of Brita water filtration products were similar to sales of PUR products (Figure 3.3), and so two extraction experiments were carried out comparing PUR and Brita faucet-mounted filters. Though the variability between the two experiments was high, the percent recoveries from the PUR filters was consistent with previous recoveries using the same extractant (in Table 3.8, for example, compare PUR results to the recoveries using 1% Peptone/1% Tween-80 extractant). Compared to PUR filters, the recoveries of the three bioagents were very similar from the Brita filters (Table 3.10). Both the PUR and the Brita filters have an outer, particulate filter surrounding an inner carbon block filter; additionally, the PUR filter has an inner, mineral-based matrix to "freshen" the taste of the water. Both the PUR and Brita filters have the same capacity (filtering 100 gallons of water).

Table 3.10
Percent recoveries of *B. subtilis* spores, *E. coli* and PP7 virus from 3-stage PUR Ultimate™ filters and 2-stage Brita On-tap™ filters. Filters were extracted with 1% peptone / 1% Tween-80 extractant. Values shown are the means of two separate experiments (Standard Deviations)

Agent	PUR Ultimate™ Filters	Brita On-Tap™ Filters
	% Recovery	
B. subtilis spores	12.3 (0.1)	15.1 (5.0)
E. coli	15.2 (10.9)	17.7 (5.4)
PP7 Virus	16.9 (2.3)	9.9 (11.1)

Microbial Sorption/Desorption from Aged PUR Ultimate™ Filters

The application of doing bioforensic analysis of household filters to identify the extent of bioagent release will typically be carried out with filters of varying age used in households. Therefore extractants must be tested with representative "used" filters of varying age. As described above, PUR Ultimate™ filters were installed on a manifold designed to pump tap water through filters at a rate and frequency representative of typical U.S. household use at the kitchen sink. With the few remaining 1-stage filters, six of these filters were installed in parallel with six, 3-stage filters and pumped 100 gallons through the filters over a week's period of time. The filters were back-flush-extracted with peptone with and without Tween-80. The filter extractions yielded similar recoveries as compared to new filters, though extraction efficiency of PP7 and spores were unusually high from the 1-stage filters (Table 3.11).

Table 3.11
Percent recoveries* of 3 organisms from 100% aged carbon block filters

	1-stage		3-stage	
Organism	1% Peptone	1% Peptone + 0.1% T80	1% Peptone	1% Peptone + 0.1% T80
PP7	52.3% (15.2)	-	11.3% (4.1)	15.9% (7.2)
E. coli K12	26.7% (6.8)	10.9% (2.4)	30.7% (4.8)	29.5% (7.9)
B. subtilis spores	-	72.2% (6.4)	10.7% (6.9)	14.2% (7.9)

*Percent recoveries are from 1 experiment using 3 replicate filters for each treatment (S.D.).

After aging the filters for 20%, 50% and 100% of their lifetime use of 100 gallons, the filters were removed from the manifold and challenged individually with virus and bacteria. After the challenge the filters were back-extracted according to the procedures described above. Though the variability was high, extraction efficiency with aged filters was at least as high as with new filters (Table 3.12). Heterotrophic bacteria were extracted along with the target microbes used to challenge the filters, but due to the selective or differential media used, the target microbes were detectable. It can be concluded that no matter the age of the filter, surrogate pathogens can be extracted and enumerated in the presence of the typical background heterotrophic bacteria.

Table 3.12
Percent extraction efficiency from filters new out of the package (after being pre-wetted), and filters that were exposed to tap water for 20%, 50% and 100% of their volume limit of 100 gallons

Organism	New Filters	20% aged	50% aged	100% aged
E. coli K12	17.6% (10.3) n=9	25.8% (3.2) n=2	44.1% (13.5) n=2	29.6% (7.9) n=3
B. subtilis spores	11.0% (5.1) n=11	18.2% (0.7) n=2	8.7% (1.6) n=2	14.1% (7.9) n=3
PP7 Virus	14.9% (0.4) n=11	5.1% (1.1) n=2	28.8% (15.4) n=2	15.9% (7.2) n=3
MS2 Virus	Not Done	5.8 (1.1) n=2	6.2 (0.5) n=3	Not done

All experiments were conducted in the presence of 0.1% Tween 80 and 1% peptone using 3–stage filters. Three separate experiments were run with the new filters, and the other experiments were done once, with n representing the number of replicate filters used.

Summary of Work to Identify an Optimal Extractant and Proposed Extraction Method

To summarize efforts to develop an optimal microbial extraction method, results from experiments done with PAC were helpful in predicting the behavior of bacterial interactions with carbon block filters, whereas viral results were quite different between PAC and carbon block media. From experiments done with new, out-of-the-package filters, it was determined that the presence of peptone and a surfactant such as Tween-80 were required for optimal extraction of representative vegetative (*E. coli*) and spore (*B. subtilis*) forms of bacteria, and this extractant functioned to extract PP7 virus. Surprisingly, varying the pH between 5 and 9 did not result in very obvious effects, with a pH of 7 typically the optimal. Related to this was a recent test in the presence of 1% Peptone and 1% Tween-80 in which we varied the concentration of MOPS buffer from 0 to 50 to 100 mM; % recovery of all virus or bacteria were unaffected, and so we have since omitted using MOPS (data not shown). Lastly, though the age of the filter affected recovery efficiencies, the trends were very similar.

Therefore, in an effort to integrate this multivariate effort to identify the optimal extractant, as well as to keep the extractant as simple and utilizable as possible, we propose the use of an extractant of 1% peptone and 1% Tween-80 which extracts consistently the highest levels of vegetative and spore forms of bacteria, as well as virus. The native pH of this solution is typically 6.8 – 7, a pH range that optimally extracts bacteria and virus and so no pH'ing is necessary.

Shown on the following two pages are the protocols recommended for extracting the two most popular brands of POU filters, the PUR Ultimate™ and the Brita On-tap™ filters.

Proposed Standard Operating Procedure: Microbial Extraction of PUR or Brita Faucet-mounted Filters.

Equipment and Supplies.
Peristaltic Pump (Eg., Cole Parmer Cat. # 07520-40) with Pump Head (#7519-06)
2 meters of Pump Tubing per Filter Extracted (PharMed size 17, Cole Parmer Cat. #06485-17)
Ring Stands
Modified PUR or Brita Filter Casings – Reusable (see next page for protocol to modify and attach barbed hose fittings to inlet and outlet of filters).
Extractant: 1% Peptone / 1% Tween-80 (sterilized by 30 min autoclaving @ 120° C, 20 psi). Need 0.5 L per filter to be extracted.
1L sterile containers (Eg. 1L Ehrlenmyer flasks with foam plug) (need one per filter)

Procedure.

1. Remove filters from casings and insert in casings modified with barbed hose fittings. Typically, 2 filters are extracted simultaneously, but more can be extracted depending on the peristaltic pump and pump head used.
2. With enough length of tubing to pump the extractant into the filter (1 m.), clamp the tubing onto the filter outlet fitting and place other end of tubing into extractant. Due to vibration from pumping, the tubing must be held in place in the collection container by a foam plug (or other means of preventing loss of sample).
3. Slip another length of tubing (1 m.) onto the filter's inlet and place end in sterile 1 L container to collect the filter extract. Clamp filters on ring stand.
4. Back-flush filters by pumping extractant at 0.5L / min through the filter's outlet, and collect filtrate into individual sterile containers.
5. Samples can be assayed for detection directly, or, for increased bacterial and protozoan sensitivity, the 0.5 L can be concentrated by centrifugation for 30 minutes at 10,000 RPM. Virus can be concentrated by ultrafiltration.

Detection.

A. *Bacillus* spore counts by pasteurization and plating on nutrient agar. Pasteurize sample (maximum volume of 25 mL) for 20 minutes at 70°C. We have found this pasteurization protocol to eliminate 100% of *B. subtilis* vegetative cells. Plate 0.5 mL for detection limit of 2/mL.
B. *E. coli* counts by plating on m-Endo (media is selective and differential for *E. coli*). Plate 0.5 mL for detection of 2 CFU/mL.
C. *S. typhimurium* can be enumerated on the selective, differential *HE (*Hektoen Enteric) agar. Plate 0.5 mL for a detection limit of 2/mL.
D. PP7 viral counts by plating on lawn of *P. aeruginosa* on nutrient agar. Plate 1 mL for detection of 1 PFU/mL (EPA Method 1602).

Fittings and Modifications for PUR or Brita Vertical Filters to Allow Extraction

Note: in order to back-flow elute microbes from PUR or Brita faucet-mounted filters, the filter effluent port was modified to allow a hose barb to be threaded into the bottom of the filter (see "Effluent Fitting" below) and a standard barbed fitting was modified to thread into the influent faucet threads on the filter (see "Influent Fitting" below).

Effluent Fitting

1/4" hose barb - 1/8 NPT male brass fitting (McMaster Carr # 5346K13) threaded into 1/8 NPT Female brass compression tube fitting (McMaster Carr # 5272K283). Remove brass nut on compression tube fitting to expose 5/16-20 NS thread.

Drill through the outflow orifice with a "G" drill (.261" dia) (a 17/64" drill will work if letter-size drills are not available*). This will remove the plastic "cross" inside the hole and size the hole for a 5/16-20 NS size tap. After tapping the hole, install assembled brass fittings by threading the brass 5/16-20 male thread into the 5/16-20 tapped outflow orifice. Use teflon pipe tape on all threads for sealing.

(*Note: this is a non-standard thread size used primarily in compression fittings. If a tap cannot be purchased locally, it can be purchased from an industrial supplier (Eg., MSC, part # 74841149).

Influent Fitting

The PUR and Brita filters inflow ports are threaded to accept a 55/64"-27 male thread found only in faucet heads employing aerators. An adapter combining this thread with a hose barb needs to be machined.
A PVC 1" X 1/2" barbed reducing coupling (McMaster Carr # 48315K72) can be adapted by turning the 1" dia. barbed end to 55/64" (0.859") diameter and cutting the special thread on a lathe. Best sealing is achieved by first removing the aerator screen installed in the filter inflow and threading the modified fitting solidly against the white sealant at the back of the inflow port.

Note: Modified filter casings are re-usable after disinfection with bleach -- they cannot be autoclaved. Soak casings in fresh 6% bleach for 2 hours, rinse repeatedly with sterile water, and dry overnight at 45°C to drive off chlorine residue.

Suppliers:

McMaster Carr	MSC
(562) 692-5911	1-800-645-7270
www.mcmaster.com	www.mscdirect.com

Effect of Tap Water Chlorine on Microbial Sorption and Extraction from POU Filters

Chlorine level and number of heterotrophs were monitored in the tap water periodically over the last three years and have found, as one would expect, an inverse relationship between the two factors (Table 3.13). Data in Table 3.13 were drawn from a six month period in 2003-2004, and, though the chlorine levels were higher, the relationship between number of heterotrophs and levels of chlorine was similar in 2005 and 2006 (data not shown).

Table 3.13
Levels of chlorine and heterotrophic bacteria found in NMSU tap water between Sept. 2003 and March 2004. Samples in 2005-2006 have given similar relationships

Chlorine Level mg/L	n (Number of samples)	Heterotrophs CFU / L
0 – 0.1	29	141,000
0.1 – 0.3	4	2,350
0.3 – 1.3	9	1,520

The experiments described above examining sorption and extraction from PAC and filters were carried out using autoclaved tap water which had little or no chlorine left after sterilization (chlorine levels ranged from below detection up to 0.02 mg/L). The last experiments performed for this project were those in which representative microbes were spiked into untreated tap water, and were distributed onto POU filters on a manifold designed to represent household water use (see the "Mock Bioagent Release Experiments" below). During these mock bioagent release experiments, the tap water used ranged in chlorine levels from 0.03 to 1.3 mg/L. After two bioagent release experiments in which no spiked *E. coli* or PP7 virus were detectable (and yet *B. subtilis* spores were traceable), experiments were carried out to test whether it was the chlorine in the tap water that was rapidly inactivating *E. coli* and PP7. Data in Table 3.14 demonstrate that *B. subtilis* spores are recoverable independent of chlorine level, whereas *E. coli* and PP7 virus were recoverable only when the chlorine levels were lowered by autoclaving or by the use of RO water. Though the tests run here were not perfectly controlled (ie., at least two factors, chlorine level and presence of native microbiota, were varied in the experiments), a thorough study to verify what is already well-documented in the literature as to the effects of chlorine on enteric microbes was not considered necessary.

Table 3.14
Survival of three bioagents in the presence of different sources of water. Microorganisms were inoculated at 10^2/mL into 1 L, and the matrices were sampled after 10 minutes to determine bacterial or viral concentration. The inocula were quantified before being added to the matrices. This experiment was repeated three more times with similar results (data not shown)

Matrix	Chlorine --mg/L--	*B. subtilis* spores	*E. coli*	PP7 Virus
		---------	% of Inoculum Recovered	--------
Non-Sterile Tap Water	0.22	82.3	0	0
Sterile Tap Water	0.02	94.6	46.9	0
R.O.* Water	0.00	82.3	28.1	35.9

* R.O. = reverse osmosis treated tap water

Mock Bioagent Release Experiments

As described above in the methods section, *B. subtilis* spores, *E coli* vegetative cells and PP7 virus were spiked into tap water and pumped through PUR Ultimate filters that had been aged before the bioagent release to 50% of their capacity. (*Salmonella typhimurium* instead of *E. coli* was used in the experiments reported in Table 3.16.) The microbes were spiked into 15 L of tap water, at a concentration of 10^2 / mL, and this inoculum was pumped onto 6 filters at a flow rate typical of American household use (0.8L/min). The filters were then exposed to tap water on the regularly scheduled basis used to age the filters. Four hours, 2-days and 5-days after releasing the microbial agents, two filters were removed from the manifold and extracted with 1% peptone and 1% Tween-80 according to the standard operating extraction procedure described above.

The levels of chlorine in the tap water used in the mock release experiments ranged from 0.03 to 1.3 mg/L, and in 5 of 5 experiments *E. coli* was inactivated after a 10-minute exposure to tap water. During the majority of timepoints assayed throughout the five day experiments, the chlorine levels were high, averaging 0.38 mg/L. No *E. coli* was detected by agar plate counting from extractions of any of the filters. In 4 of 5 experiments, likewise there was no PP7 virus detectable after 10 minute exposure to the tap water. In the one experiment in which PP7 did survive the initial exposure to tap water, 4% if the initial number calculated to have been retained on the filter was extracted four hours after the challenge; in this experiment, no PP7 virus was detectable after 2 days nor 5 days after the challenge. Experiments were performed to verify that *E. coli* and PP7 virus are highly sensitive to the levels of chlorine typically observed in these mock challenge experiments (see Table 3.14), and so we feel confident in concluding that *E. coli* and PP7 virus were rapidly inactivated in the tap water and so were not recovered from filters due to this inactivation.

Similar to single-filter challenges, *B. subtilis* spores were retained by the filters in high numbers ($\geq$ 99% of the numbers challenged). Subsequent effluent samples on the same day of the challenge showed very little spore counts (less than 1% of the initial challenge), as did daily samples of the effluent. Extraction of spores from filters 4 hours after inoculation gave an average recovery of 10.4% of those calculated to have been retained on the filters. The number of extracted spores decreased two days and five days after the initial challenge to 6.4 and 4.3% of the spores retained on the filters (Figure 3.5 and Table 3.15). It is interesting to note that our technique to monitor spores in this mock-release experiment was selective for spores only, due to our use of pasteurization to enumerate *B. subtilis* spores. When the vegetative form of *B. subtilis* was assayed for by plate counts in these experiments, no *B. subtilis* cells were found, indicating that the bacillus spores did not germinate while retained on the filters.

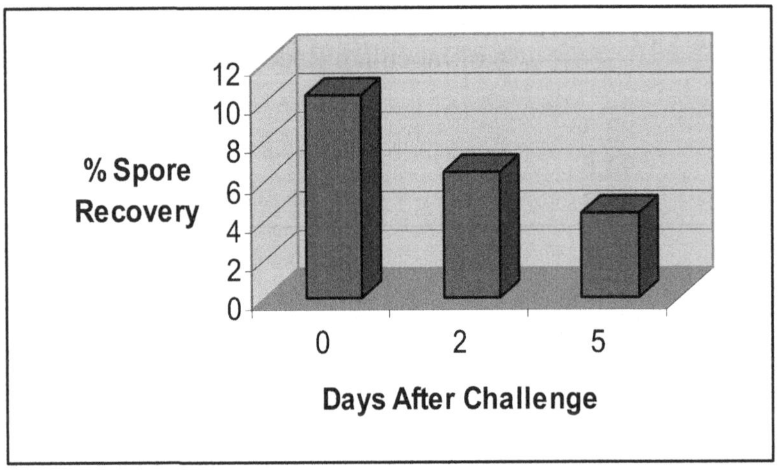

Figure 3.5. Percent extraction efficiency after a mock release of *B. subtilis* spores. The graph shows the percent of the initial population retained by the filters that were extracted after 4 hours (Day 0), 2 and 5 days after the release event. After the initial challenge, the filters were exposed to tap water on a schedule that reflected typical U.S. household water use at the kitchen sink. Three-stage PUR filters that were 50% aged before spore exposure were used.

Table 3.15

Results from individual mock-release experiments showing the percent of *B. subtilis* spores recovered from duplicate PUR filters 4, 52 and 124 hours (0, 2 and 5 days) after the initial challenge

	% Spore Recovery After:		
Date	4 hours	52 hours	124 hours
7/6/2005	3.3	3	1.7
9/28/2005	16.5	5.2	7
10/12/2005	23.8	4.6	1.5
10/26/2005	1.6	1	0.7
12/7/2005	6.7	18.3	10.6
Mean (n=5)	10.4	6.4	4.3
Std.Dev.	9.5	6.8	4.3

The last experiments performed were designed to test whether challenged filters could be stored at 4°C for three days without affecting detection ability. Additionally, in addition to challenging the filters with *B. subtilis* spores and PP7 virus, we utilized *Salmonella typhimurium* instead of *E. coli*. As observed in previous experiments, the only organism that was consistently recovered from the filters were the *B subtilis* spores, the other enteric agents evidently inactivated by the typical levels of chlorine (0.3 – 0.4 ppm). Two days after the mock release event, four filters were removed from the manifold, and two were extracted immediately and two were refridgerated for 3 days before extraction. In both experiments, day-0 and day-2 recoveries of spores were within the range of the recoveries reported previously (compare Tables 3.15 and 3.16). Spores were still detectable when extracted from the stored filters, and were extracted at efficiencies similar to filters that were on the tapwater manifold for five days (Table 3.16). This result showing that filters do not have to be extracted immediately in order to detect filter-retained spores obviously makes the proposed approach more feasible.

Table 3.16
Percent *B. subtilis* spore or *Salmonella typhimurium* recovery 4 hours after mock release (Day 0), and 2 days after the release. On day 2, duplicate filters were also stored at 4°C for 3 days, after which they were extracted as on day-0 and day-2. Fifty percent-aged filters were used and tap water-flow patterns were as in the challenge experiments described in Table 3.15 and Figure 3.5. (Standard deviations are shown in parentheses)

	% Microbe Recovery After		
Date	Day 0	Day 2	Stored @ 4°C for 3 days
2/1/2006 (Spores)	15.8 (0.1)	11.3 (2.3)	9.7 (0)
2/1/2006 (Salmonella)	0	0	0
2/15/2006 (Spores)	3.8 (0.1)	1.3 (1.1)	0.8 (0)
2/15/2006 (Salmonella)	0.5 (0.7)	0	0

These results showing that a spore-release event can be detected up to five days after it occurred lend credence to the idea of using existing POU filters as bioforensic devices to aid in the analysis of a bioagent release event should it occur. Additionally, data shows that it is possible to store filters for three days at 4°C and extracted spores are still detectable. Overall, the extraction efficiencies of viral and bacterial agents was low (ie., less than 50% of those attached), indicating that the microbes are tightly adhered or associated with the filter matrices. The data obtained in this study did not require post-extraction concentration, and so, though the percent extraction efficiency was low, steps can be taken to concentrate bacteria and virus (by centrifugation and ultrafiltration, respectively) when detecting lower levels of agent. In non-chlorinated water supplies, it is probable that other microbial agents than spores would survive in water distribution systems, and thus be amenable to detection by these methods (We were able to extract vegetative bacteria and virus when non-chlorinated water was used – see data above.) We would thus conclude that, using the methods proposed here, microbial extraction of household POU devices can provide an important link in bioforensic studies establishing the identity, and therefore possibly the source, of bioagents released in potential attacks on the U.S. domestic water supply.

Nucleic Acid Extraction and Analysis

Numerous nucleic acid extractions were attempted from bacteria extracted from the filters (see SOP.DNA Extract1.2 in Appendix 1). To date, though it is possible to extract and amplify DNA from *E. coli* and *Bacillus* spore control cultures (data not shown), it has not been possible to extract high quality DNA that is PCR amplifiable. DNA has been extracted in the presence of fairly high concentrations of EDTA (20 mM), which we have previously shown to prevent nuclease activity in environmental extracts, as well as DNA damage due to free hydroxyl radicals (Ruiz et al. 2000). Because activated carbon has been observed in all cell pellets extracted from filters, it is hypothesized that the DNA is sorbing to the activated carbon as we have observed before in extracting DNA from activated-carbon amended wastewater (Z. Yu, 1996 New Mexico State University PhD Dissertation). In separate DNA extraction experiments, 0.1% pyrophosphate has been added in an attempt to prevent electrostatic adherence of the DNA to the activated carbon; this resulted in the same little-to-no yield of DNA which was not amplifiable by PCR (data not shown). Additionally, a direct lysis technique was attempted in which a lysis solution was pumped (the same as that used in SOP DNA Extract1.2) onto the filter. The filters were then incubated and extracted by back-flushing. Again, the DNA was not apparent on a 1% agarose gel, nor could any DNA be amplified by PCR. In the past (Yu, 1996), we have solved this problem by adding ug quantities of t-RNA to the material before we lyse the cells/virions in order to keep DNA from binding to the activated carbon. Testing these variables to successfully extract nucleic acids, though important, is now beyond the scope of the current project.

CHAPTER 4: SUMMARY AND CONCLUSIONS

The objectives of this project were to identify the top-selling household drinking water filter and use this filter in experimentation to determine how feasible it was to use these filters as bioforensic devices in case of a bioterrorism release in U.S. drinking water sources. Collaborative work with the Water Quality Association resulted in modifying a national survey to include the necessary information regarding the most commonly used Point-of-Use filter used in U.S. households. Survey results indicated that PUR faucet-mounted filters were the most commonly used, and follow-up work with American Tank Company demonstrated that of all the PUR water filter products, the PUR Ultimate™ filter was the top-selling POU device, and so, was the filter used in this project.

It was found that powdered activated carbon (PAC) and carbon block filters retained high percentages of virus and bacteria. Carbon block filters (either 1-stage or 3-stage PUR Ultimate™ filters) retained greater that 99% of spore and vegetative forms of bacteria, whereas virus were retained to a lesser degree by the filters (76 % $\pm$ 8% were retained on 1-stage filters and 92% $\pm$ 5% were retained on 3-stage filters (+/- S.D.). After increasingly complex experimentation with PAC and PUR carbon block filters, it was determined that a combination of proteinaceous (1% peptone) and surfactant (1% Tween-80) components were optimal in extracting virus and bacteria from PUR Ultimate™ faucet-mounted filters.

A back-flow elution protocol was chosen to extract microbes from POU filters for the following reasons: 1. Due to the 1 um pore size of the carbon block used in the two top-selling POU filters, the majority of bacteria should be concentrated on the outer surface of the carbon block, and so eluting bacteria via back-flushing would minimize exposure to the highly sorptive activated carbon in the interior of the carbon block. 2. Access to carbon blocks in the interior of both PUR and Brita filters is difficult, requiring labor-intensive sawing and prying to remove the carbon block. 3. Though the approaches are not directly comparable, viral recoveries were at least an order of magnitude higher when viruses were eluted from intact filters compared to when isolated PAC was homogenized and chemically extracted. Using the proposed POU filter elution method, typically, 15 - 20% of attached virus or bacteria could be extracted from the filters, the age of the filter had little influence on extraction efficiency, and recoveries were similar from the two top-selling faucet-mounted filters. Thus, though the extraction efficiency is relatively low, the method is robust in that it extracts representative vegetative and sporulated bacteria and representative virus from filters of varying age and different manufacturers in a consistent manner.

In repeated mock bioagent release experiments (n=5), *Bacillus subtilis* spores were detected in filter extracts in each of the mock challenges up to five days after the release event (five days was the maximum time tested). In these mock release experiments, representative enteric agents (*E. col, S. typhimuriumi* and PP7 virus) were inactivated within 10 minutes of exposure to the tap water, which had levels of chlorine averaging 0.38 mg/L. These types of microbes would therefore not be of great concern in regard to protecting chlorinated U.S. water supplies, but obviously non-chlorinated supplies, or chlorine-attenuated supply sections, would probably be vulnerable. Results demonstrate that household POU filters can be effectively used in bioforensic applications to survey the extent and potential origin of a bioagent release in U.S. domestic water supplies.

It is of concern that, due to the fairly low microbial extraction efficiencies from faucet-mounted filters, it would be technically difficult to detect low concentrations of filter-associated bioagents. No post-extraction concentration methods were used in this study, and it is recommended that such methods be adapted to this filter-extraction approach. Centrifugation and/or ultrafiltration would be fairly straight-forward methods to concentrate viral, bacterial and protozoan agents. An ultrafiltration method in which virus, bacteria and protozoa are simultaneously concentrated from turbid surface waters has been developed (Morales et al. 2003), and this would be a reasonable first step to take. Related to detection methodology, is the continued need to develop a molecular assay for detecting agents extracted from carbon-based filters. The co-elluting presence of activated carbon is probably the most important factor that has prevented nucleic acid isolation and PCR amplification. Extracted nucleic acid and activated carbon are probably tightly associated and so efforts to separate the two are needed.

APPENDIX A: STANDARD OPERATING PROCEDURES

SOP.Carbon.1.4. Sorb and Extract Known Cell/Viral Quantities from PAC.

Materials: Nutrient Broth (NB), 10 mM MOPS pH 7, centrifuge tubes, 1.5 ml micro centrifuge tubes, 0.85% saline, NA or TSA plates, soft agar (for plaque assays).

Day 1: Inoculation

Bacteria- inoculate 3 mL of NB with a single colony, shake (150rpm) at 37 C for 17 hrs.

Bacillus spores- use 6-day spore prep stored at 4° C in 40% ethanol.

PP7- inoculate 35 ml of NB from *P. aeruginosa* slant, incubate overnight in a 37 C incubator, not shaken - next day, add 1 ml of overnight into 35 ml of NB and incubate at 37° C for four hours (for use in PP7 virus plaque count plates). On Day 2, to plate virus: Add 1000 uL virus sample and 500 uL 4-hr *Pseudomonas* culture into 4.5 mL soft agar (SA) and vortex, then pour on plate and let dry for 30 minutes before inverting and placing in incubator (using EPA 1602 protocol as basis for viral enumeration).

Day 2: Adsorption

1. Add 10 ml of 10 mM MOPS pH 7 to centrifuge/homogenizer tubes (3 tubes per treatment)
2. Add culture suspended in saline to a concentration of 10^6 cells / mL (i.e. 10^7 total)
   -for *E. coli* and *P. aeruginosa* centrifuge 1 ml for 2 min @14000 rpm, decant supernatant, and add 1 ml of saline, add 10 uL to each tube
3. Vortex each tube
4. Take an input sample of 100 uL from 3 tubes, this will be the 10^-1 dilution, dilute out until 10^-3 and plate the 10^-3 sample, 2 replicates of 25 uL
5. Add 1 gram of carbon to each tube
6. Set tubes on shaker at room temp for 30 minutes
7. Centrifuge tubes for 3 minutes at 3000 rpm
8. Take 100 uL samples from 3 tubes, this will be the unattached numbers, plate out -1 through -3 dilutions
9. Decant supernatant

Desorption:

1. Keep samples and solutions on ice
2. Add 10 ml of treatment solution to each tube just before homogenization
3. Homogenize sample at 8000 rpm for 1 minute in an ice bath
4. In between samples clean blade with a 30 second spin of 80% ethanol (5% bleach for spores), and 2 water spins for 30 seconds each
5. Every 3 samples use fresh ethanol and sterile water to rinse
6. Centrifuge samples at 3000 rpm for 3 minutes
7. Take desorped sample from each tube (100 uL= 10^-1 dilution)
8. Plate out -1 through -3 dilutions

Day 3: Enumeration

1. Count plates ~ 17 hours after incubation
2. Use counts between 5-50 colonies when possible
3. Determine average CFU/ml for each treatment, standard deviation, and percent recovery (average CFU/ml/attached (input-unattached) x 100).

SOP.Filter.1.2.
Sorb and Extract Known Quantities of Cells/Virus from POU Filters.

Materials:
6- 1000 mL sterile collection flasks
6- 1000 mL flask for
12- 500 mL flasks
6- Masterflex #17 C-flex tubing
3- clamps
3 PUR filter cases
6 PUR filters
1 M MOPS pH 7
3 L sterile tap water
6 L sterile RO water
Nutrient broth
Nutrient agar plates
TSA plates
Soft agar (8.2 g/L)
0. 85 % Saline
P. aeruginosa slant
50 mL falcon tubes
1.5 and 23 mL eppendorf tubes
virus tubes

Day 1:
- Bacterial and viral inoculations done the same as described in SOP.Carbon.1.4
- Bleach and rinse filter cases and place in 40- 50°C incubator overnight

Day 2:
Prepare Inoculant:
1. Inoculate 50 mL NB with 1 mL of *P. aeruginosa* overnight culture and place in 37 C incubator for 4 hours (for use in PP7 plaque count plates).
2. Make 4.5 mL SA tubes (2 per dilution) and 2.7 mL NB tubes (1 per dilution)
3. Prepare autoclaved extractant--typically 1% peptone/1%Tween-80
4. Pellet 1 mL of *E. coli* K12 cells and resuspend in 1 mL saline, dilute 1:200
5. PP7 virus stock should be diluted to 10^ -5 and a total volume of 3 mL
6. Inoculate 3 L sterile tap water with 3 mL PP7 (10^7/L)and 600 uL *E. coli* K12 (10^6/L)
7. Take input sample and pour 500 mL inoculated water into 6 flasks

SOP.Filter.1.2. (Continued)

Challenge Filters:

1. Attach and clamp tubing to influent and effluent barbed fittings.
2. Set filter pump to 3 (0.5 L per minute)
3. If filters are new, pump1 L of sterile RO water through 3 filters
4. Let pump run air through until the filter until it stops dripping water
5. Run 500 mL of inoculated water through filters collecting the filtrate and sample for unattached organisms. Continue pumping until liquid no longer comes out the filter effluent line.
6. After all filters have been inoculated, place pump on tubing leading to effluent port to allow for back-flushing of the system.
7. Back-flush filters with treatment solutions and collect filtrate to sample for desorped organisms.
8. If concentration is needed, centrifuge extracts at 10,000 rpm for 30 minutes, decant supernatent and add 0.5 mL 0.1 % Tween 80 + saline, vortex and dilute for plating.

Plating

- Use drop plate for *E. coli* K12 samples
- Plate undiluted and 1:10 diluted samples of input
- Plate –1 through –3 of pelleted desorped sample

Use pour plate overlay agar method for virus

- Plate –2 through –4 for input
- Plate –1 and –2 unattached
- Plate –1 through –3 of desorped

Place all plates in 37 C incubator for 15-20 hours

Day 3:

- Count plates and determine % recoveries using the following formulae:

[1] No. of attached microbes = No. before filtration – No. unattached after filtration

[2] % Extraction Efficiency = No. of extracted microbes / No. attached microbes x 100

SOP.Manifold Challenge.1.3.

1. Let the filters run twice on the day of the challenge with normal tap water
2. Aim for $1x10E^2$/mL, and since there are 15 L in carboy, inoculate with 1.5 X 10E^6 cells of *E. coli*, *Bacillus* spores and MS2 (or PP7) virus, then stir the sample well. Drain the influent line of 250 mL, then take an input sample (need about 4 mL).
3. With the faucet off, allow the pump to run at the normal time and allow the tilted carboy to drain as much of the inoculum as possible. Before air is pumped, turn on faucet and allow the system to run the rest of the 5 minute cycle.
4. Take a sample of the effluent from the system after one minute, and after 4 min. These effluent samples represent unattached cells and allow us to calculate the average number of attached.
5. For the next four cycles take a sample of the effluent after 2 minuets into each cycle. (i.e. if you start the cycles at 10:30 than you take the two minute samples at 11:32, 12:32, 1:32 and 2:32)
6. After the fourth pump (at ca. 2:35 PM) run, remove the two outermost filters on the manifold and follow the normal filter extraction procedure.
7. The next day (day 1), take an effluent sample in the morning (at 10:32) and in the afternoon (at 2:32).
8. At day two (Friday), take another effluent sample and then extract the next outermost filters after running two cycles on that day and follow the normal extraction procedures. Repeat this step on the fifth day.

Disinfecting the manifold

1. Tape the filter holders closed and run 10 L of 0.5% bleach (1:10 of commercial bleach) through the filter holders.
2. Let the manifold sit for two hours and than run 10 L of freshly diluted (1:10) bleach through the system.
3. Allow the system to sit for two hours.
4. Than remove the carboy and rinse out well, than re-attach the carboy to the system and allow tap water to run through the system for two days.
5. After two days measure the chlorine and spore levels. No *B. subitlis* spores should be present, and chlorine levels should be back to normal levels (ie., 0.03 – 1.3 mg/L).

SOP Spore1.1

Method for the purification of *Bacillus subtilis* var *niger* endospore purification
V. Hill Lab (Centers for Disease Control)

MCCB SOP V-07 from USEPA was used as a guide

1. *B. subtilis* was grown at 37°C for 6 days in generic spore media (GSM). Recipe for GSM was followed explicitly except $MnSO_4$ was substituted for $MnCl_2$.
2. On the 6th day the culture was evenly split into 15 ml conical tubes and placed in a 70°C water bath for 20 minutes.
3. Tubes were centrifuged at 4000 g for 20 minutes.
4. Supernatant was removed and 10ml of cold, sterile dI water was added. The tubes were vortexed to mix and centrifuged at 4000g for 10 minutes.
5. Step 4 was repeated but tubes were centrifuged for 20 minutes.
6. Supernatant was removed and pellets were resuspended in 40% ethanol and vortexed.
7. 1 ml aliquots were placed in 2ml cryotubes and are stored at 4°C.

Malachite green staining

1. Place 10 ul of spore suspension on slide and allow to air dry (approx 30 minutes)
2. Place slide on staining rack over a beaker of boiling water and flood with Malachite Green 5%
3. Let sit for 10 minutes
4. Rinse slide until water runs clear
5. Flood slide with safranin
6. Let sit for 1 minute
7. Rinse until water runs clear
8. Air dry and observe under oil immersion

SOP.DNA Extract1.2 DNA Extraction from Cells Extracted from POU Devices.

Needed:
Lysis buffer: 50 mM TRIS, 20 mM EDTA 1% SDS with 50 ug/mL Proteinase K (pH8).
Sterile microfuge tubes (1.5 mL and 2 mL).
Sterile bead mill tubes (with 1 g of 0.1 mm beads).
A Bead Mill.
TE Buffer 10 mM TRIS 1 mM EDTA (pH 8)

1. After microbes have been extracted from POU filters and samples taken for cell counts, place 200 mL in each of 2 sterile centrifuge bottles and spin at 10,000 RPM for 30 minutes. Carefully siphon off the supernatant, leaving about 2 mL in the bottle.
2. With a 1 mL pipetor rinse off the walls of bottle, making sure to collect all adhered cells. Transfer into 2 mL microfuge tubes and spin for 3 minutes, 14,000 rpm. Decant supernatant, and resuspend cells of one of the tubes in 0.5 mL of lysis buffer, and add to other pellets (to "pool" the sample back into 1 tube).
3. Incubate sample in lysis buffer 1 hour at 37 deg. C, vortexing every 15 minutes.
4. IF tracking spore lysis, withdraw 10 uL to spore stain to examine extent of lysis.
5. After incubation, add 0.5 mL of lysate to sterile bead mill tubes and bead mill at high setting (4300 rpm) for 20 seconds.
6. Pellet cell debris for 2 minutes at 10,000 rpm and withdraw supernatant to new 2 mL microfuge tube. (If needed, withdraw 10 uL for spore stain.)
7. Rinse beads with 0.5 mL TE and add to collected supernatant.
8. Precipitate with 0.1 vol of NaAcetate and 2 vol of 100% EtOH.
9. Spin at 10,000 rpm for 5 minutes and then decant supernatant.
10. Wash the pellet by adding 1 mL of cold 70% EtOH, vortex, and spin at 10,000 rpm for 2 minutes.
11. Decant supernatant, and allow pellet to dry, (Speedvac if needed) and resuspend in 10 - 50 uL TE, depending on expected yield.
12. Store at 4°C in fridge. Run out on agarose gel to visualize the DNA yield, and quantify yield by measuring 260/280 nm absorption.

REFERENCES

Bell, F.A., Jr., D.L. Perry and J.K. Smith. 1984. Studies on home water treatment systems. J. Amer. Water Works Assoc. 76:126-130.

Brita on-line store, 2006. [online]. http://www.shopbrita.com/index.asp Cited April 15, 2006.

Camper, A.K., M.W. LeChevallier, S.C Broadway and G.A. McFeters. 1985a. Evaluation of procedures to desorb bacteria from granular activated carbon. J. Microbiol. Methods 3:187-198.

Camper, A.K., M.W. LeChevallier, S.C Broadway and G.A. McFeters. 1985b. Growth and persistence of pathogens on granular activated carbon filters. Appl. Environ. Microbiol. 50:1378-1382.

Chaidez C., and C.P. Gerba. 2004. Comparison of the microbiologic quality of point-of-use-treated water and tap water. Inter. J. Environ. Health Res. 14:253-260.

Garduño, R.A., E. Garduño, M. Hiltz, and P. S. Hoffman. 2002. Intracellular growth of *Legionella pneumophila* gives rise to a differentiated form dissimilar to stationary-phase forms. Infect. Immun. 70:6273-6283.

Gauthier, A. M. de Grado, and B. B. Finlay, 2000. Mechanical fractionation reveals structural requirements for enteropathogenic *Escherichia coli* Tir insertion into host membranes. Infect. Immun. 68:4344 - 4348.

Han, I., M.A. Schlautman, and B. Batchelor. 2000. Removal of hexavalent chromium from groundwater by granular activated carbon. Water Environ. Res. 72: 29-39.

LeChevallier, M.W. and G.A. McFeters. 1985. Interactions between heterotrophic plate count bacteria and coliform organisms. Appl. Environ. Microbiol.49:1338-1341.

Mayer, P., W. DeOreo, E. Opitz, J. Kiefer, W. Davis, B. Dziegielewski, and J.Nelson. 1999. Residential End Uses of Water. AWWA Research Foundation and the American Water Works Association.

Morales-Morales, H.A. G. Vidal, J.Olszewski, C.M. Rock, D. Dasgupta, K.H.Oshima, and G. B. Smith. Optimization of a reusable hollow fiber ultrafilter for the simultaneous concentration of enteric bacteria, protozoa and viruses from water. Applied & Environ. Microbiol. 69:4098-4102. (2003).

New Mexico State University, 2005. [Online] http://www.ppd.nmsu.edu/pdf/Domestic_Water_Conf_Report.pdf [cited March 14, 2006]

Prescott, L.M., J.P. Harley and D.A. Klein. 2002. Microbiology. Fifth Edition. McGraw-Hill, New York.

Reasoner, D.J., J.C. Blannon, and E.E. Geldreich. 1987. Microbiological characteristics of third-faucet point-of-use devices. J. Am. Water Works Assoc. 79:60-66.

Ruiz, T.R., S. Andrews and G.B. Smith. 2000 Identification and characterization of DNA nuclease activities in anaerobic environmental samples. Canad. J. Microbiol. 46:736-740.

Snyder, J.W., Jr., C.N. Mains, R.E. Anderson and G.K. Bissonette. 1995. Effect of point-of-use, activated carbon filters on the bacteriological quality of rural groundwater supplies. Appl. Environ. Microbiol. 61:4291-4295.

Stewart, M.H., R.L. Wolfe and E.G. Means. 1990. Assessment of the bacteriological activity associated with granular activated carbon treatment of drinking water. Appl. Environ. Microbiol. 56:3822-3829.

U.S. Census Bureau. 2006. [Online] http://quickfacts.census.gov/qfd/states/00000.html, [cited March 14, 2006]

Wallis, C., C.H. Stagg and J.L. Melnick. 1974. The hazards of incorporating charcoal filters into domestic water systems. Water Res. 8:111-113.

Water Quality Association. 2004. National Consumer Water Quality-Research Summary. Applied Research West., Inc.

Watertank Literature [Online].) http://www.watertanks.com/category/328/. [cited march 14, 2006]

ABBREVIATIONS

°C	degrees Celsius
CFU	colony-forming units
DNA	deoxyribonucleic acid
EDTA	ethylenediaminetetraacetic acid
GAC	granular activated carbon
g	gram (or g force)
Hrs	hours
HE	Hektoen Enteric (medium selective for *Salmonella* species)
L	liter
m	meter
mEndo	medium selective for coliform bacteria
uL	microliter
mg	miligram
mL	milliliter
mm	milimeter
mM	milimolar
MOPS	4-morpholinepropanesulfonic acid (pH buffering reagent)
nm	nanometer
NB	nutrient broth (non-selective bacterial growth medium)
NMSU	New Mexico State University
PAC	powdered activated carbon
PFU	plaque-forming units
pH	negative log of hydrogen ion concentration
PhD	doctor of philsophy
POU	point-of-use
ppm	parts per million (mg/L)
psi	pounds per square inch
rpm	revolutions per minute
SDS	Sodium dodecylsulfate (an ionic detergent)
TE	TRIS/EDTA buffer
™	Trademark
TRIS	Tris(hydroxymethyl)aminomethane hydrochloride (pH buffering reagent)

TSB	tryptic soy agar (non-selective bacterial growth medium)
Tween-80	Polyethlene glycol sorbitan monooleate (Polysorbate-80, a non-ionic detergent
Triton X-100	4-(1,1,3-Tetramethylbutyl)phenyl-polyethylene glycol (a non-ionic detergent)

IWA Publishing's authorised EU representative for General Product Safety Regulations is Diane D'Arras, 15 rue Duret, 75116 Paris, France, e-mail: safety@iwap.co.uk.

Printed and bound by CPI Group (UK) Ltd, Croydon, CR0 4YY
17/09/2026
02218793-0011